U0392388

Education
92

大师与超级大师

Masters and Grandmasters

Gunter Pauli

[比] 冈特·鲍利 著

[哥伦] 凯瑟琳娜·巴赫 绘

田 烁 王菁菁 译

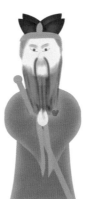

上海远东出版社

丛书编委会

主　任：田成川

副主任：何家振　闫世东　林　玉

委　员：李原原　翟致信　靳增江　史国鹏　梁雅丽

任泽林　陈　卫　薛　梅　王　岢　郑循如

彭　勇　王梦雨

特别感谢以下热心人士对童书工作的支持：

匡志强　宋小华　解　东　厉　云　李　婧　庞英元

李　阳　刘　丹　冯家宝　熊彩虹　罗淑怡　旷　婉

杨　荣　刘学振　何圣霖　廖清州　谭燕宁　王　征

李　杰　韦小宏　欧　亮　陈强林　陈　果　寿颖慧

罗　佳　傅　俊　白永喆　戴　虹

目录

Contents

两名学生正在担心自己的考试。他们要求严格的教授说了，如果谁能向他提出一个专业领域内他答不出来的问题，就能在考试中得到满分。

"我觉得这是在骗人。"一名学生说。

Two students are worrying about their exams. Their tough professor proposed that those who could ask him a question about their subject field that he cannot answer properly will get full marks in the exam.

"I think this is a trick," says the one student.

我觉得这是在骗人

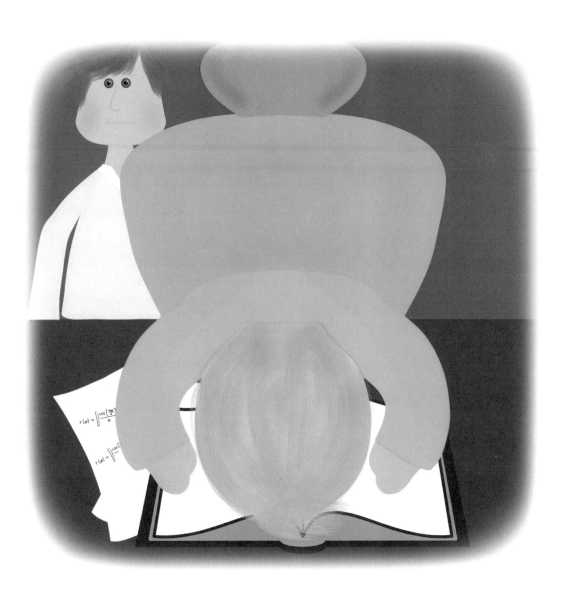

I think this is a trick

激励我们学得更好、学得更多的一种方式

A way to challenge us to learn better and more

"不，不是！这是激励我们学得更好、学得更多的一种方式。"另一位学生答道。

"但这是为什么呢？"

"因为，如果我们想提出一个他都答不出来的问题，我们首先必须要学会他所知道的一切。"

"No, it's not! It's a way to challenge us to learn better and more," replies the other student.

"But why?"

"Because if we want to ask him a question he has no clue about, we must first know everything he knows."

"这就是说，要想得到满分，我们必须像他一样学识渊博。算了吧！我们还是用心学习他教过我们的那些东西吧！"

"对他来说，坐在那里听学生们一遍又一遍地讲述他知道的一切，那得多无聊啊！"

"That means to get full marks, we must be as smart as he is. Forget it! We'd better learn everything he'd taught us by heart."

"How boring it must be for him to sit there and listen to students telling him everything he already knows over and over again."

对他来说那得多无聊啊！

How boring it must be for him!

就能让他进行思考了

It will make him think

"没错。学到一些专业领域内闻所未闻的事情，对他来说一定会非常振奋！"

"所以你看，如果我们提出一个他答不出来的问题，就能让他进行思考了。"

"Exactly. So how refreshing it must be for him to learn something about his specialty that he had never heard before."

"So you see, if we ask him a question he cannot answer, it will make him think."

"这位教授还真是与众不同。我向你保证，这就是为什么他被看作大师——一位真正的导师。"

"设想你是一位大师，每年都有数百名学生向你展现从你这里学到的东西。看到自己启发了他们，一定是一件非常美妙的事情！"

"This professor is different. I guarantee you. That's why he is considered a master – a true sensei."

"Imagine you are a master and every year hundreds of students show you that they've learned something from you. It must be wonderful to see how you've inspired them ..."

一位真正的导师

A true sensei

一群青出于蓝而胜于蓝的学生

students who will become better

"而且，这也会给大师一些启发！"

"再设想一下，如果每年都会有一些学生提出新想法、新理论、新案例，这对大师来说意味着什么？正因为他是最好的教授，所以他还能从学生那里学到东西。"

"据说，最好的大师都有一群青出于蓝而胜于蓝的学生。"

"And that will inspire the master!"

"And imagine that each year there will be a few students who bring new ideas, new theories, and new cases to him. What will happen to the master? He will learn from his students thanks to having been the best professor!"

"They say that the best masters have students who will become better than they themselves have ever been."

"没错。这是激发创新、进步和创造力的唯一途径，也是开创更美好的世界所必需的！"

"你说得很对。假如我们仅仅学到了父母和老师已经知道的那些东西，对其他一无所知，那我们不会有进步！"

"That's true. And that's the only way innovation, progress, and creativity can take place. It is so much needed to make a better world!"

"You're right. Imagine if we only learnt what our parents and teachers knew and nothing more. There will be no progress!"

开创更美好的世界

To make a better world

我们的教授成长为一位超级大师

Our professor can become the grandmaster

"是的。因为大师有这么多的学生，他可以学到很多关于自己专业领域的知识，而学生们只能向这一位大师学习。"

"这就是我们的教授成长为一位超级大师的方式。"

"Yes. And since the master has so many students, he is learning a lot about his subject field, while the students are only learning from one master."

"That's how our professor can become the grandmaster."

"这也是他的学生们成长为新一代大师的方式！"

"如果一位超级大师向自己最好的、将来有可能成为一名大师甚至超级大师的学生学习，并分享自己全部所知所学，那么他就会永远地受人尊敬和怀念。"

……这仅仅是开始！……

"And that's how his students can become the new masters!"

"If the grandmaster learns from his best students, who will later become masters, and even grandmasters, and shares everything he knows and learns, then he could even be immortalised and remembered forever."

... AND IT HAS ONLY JUST BEGUN!...

... AND IT HAS ONLY JUST BEGUN! ...

Did You Know ?

你知道吗?

Nearly 40 percent of children, including adolescents, suffer from exam stress. Students get anxious before exams, have sleepless nights, and fail because they are exhausted.

将近 40% 的儿童（包括青少年）承受着考试的压力。考试前，学生会变得焦虑、失眠，最后会因为疲惫而考试失败。

Some of the best ways to overcome exam stress are to exercise and to talk with those who had to go through the same experience: parents and older brothers and sisters.

克服考试压力的最好方式是运动，以及与那些有过相同经历的人聊天，比如父母和哥哥姐姐。

The title of Grandmaster is awarded in chess, martial arts, and by guilds of arts and crafts.

国际象棋、武术和工艺美术等领域会颁发"超级大师（特级大师）"头街。

In Oriental cultures the title of Grandmaster is a honorific and does not confer rank; rather it distinguishes a person as very highly revered in his or her field.

在东方文化中，"超级大师"是一种尊称，没有等级之分，它只是表明一个人在某一专业领域极其受人尊重。

In the Orient, a term such as "teacher" is more common than "master", although they often have the same meaning. The Japanese use the term sensei, meaning "teacher", which is literally translated as "born first". The Chinese use the term shifu, which is written as a combination of the characters "teacher" and "father" and a combination of the characters "teacher" and "mentor".

在东方文化中，"老师"一词比"大师"更为常见，虽然两者的意思通常是一样的。日本人用"先生"（sensei）这个词来表达"老师"的含义，字面意思是"先出生的人"。中国人用的"师父（傅）"一词，在汉字中是"师"和"父"或"师"和"傅"的组合。

The founding president of Tanzania, Julius Nyerere, was called Mwalimu Nyerere, which means "teacher" Nyerere in Swahili.

坦桑尼亚的建国领袖朱利叶斯·尼雷尔被称为尼雷尔马里木，在斯瓦希里语中的意思就是尼雷尔导师。

Confucius is considered a grandmaster who achieved immortality. He emphasised personal and governmental morality, correctness of social relationships, justice, and sincerity. He championed strong family loyalty and respect of elders by their children.

孔子被看作一位名垂青史的超级大师。他强调个人与政府的道德准则，重视社会关系的正确性、公平和诚实，倡导强有力的忠于家族、尊重长辈的道德准则。

印度独立运动领袖莫罕达斯·卡拉姆昌德·甘地被称作"圣雄"，在梵语中的意思是"品格高尚"或"令人尊敬"。他还被称作"巴布"，在古吉拉特语中的意思是"父亲"。

Mohandas Karamchand Gandhi, the leader of the Indian independence movement, was called Mahatma, which in Sanskrit means "high-souled" or "venerable". He is also called Bapu, which means "father" in Gujarati.

想一想

你希望有这样一种考试吗——你向老师提问题，老师将根据你知道而他不知道的内容给出分数？

Would you like an exam where you can ask your teacher questions and get grades based on what you know and your teacher did not?

Would you like to be remembered by your friends for what you have said or what you have done?

你希望你由于自己的言行而被朋友们记住吗？

Are you nervous when a teacher asks you a question? Do you get nervous because you are afraid that you do not know the answer?

当老师向你提问时你紧张吗？你会因担心自己不知道答案而紧张吗？

Do you think it is boring for teachers to correct exams and to listen to information they already know?

你认为老师批改试卷、听学生讲自己已经知道的知识是件很无聊的事情吗？

Everyone gets nervous when they have to write a test or an exam. This is most likely because we do not know what the outcome will be, or perhaps because we do not know what questions will be asked.

So let's make a list of at least five different ways, other than written or oral exams, that allow us to check and demonstrate that we understand what we have learned.

　　每个人在考试时都会感到紧张。这很有可能是因为我们不知道结果如何，也可能是因为我们不知道将会被问到什么问题。

　　所以，让我们来列张清单，至少列出 5 种方式，不包括笔试或口试，来检验自己是否理解了所学的知识。

学科知识
Academic Knowledge

生物学	有其父必有其子；进化论；下一代如何学习与适应？
化 学	当人与人之间有良好的化学反应时，他们会有很好的理解和交流。
物 理	创新依靠洞察，而这不能仅通过现有的知识来获得。
工程学	工程学的创新需要足以质疑专家的能力与信心；只有学生超越老师，才会有进步和创新。
经济学	用产生问题的逻辑不能想出解决问题的方案。
伦理学	不耻下问的谦卑态度。
历 史	讲故事在学习中的重要性；苏格拉底被看作欧洲古代史上最好的老师，他教学生思考的方法不是提供答案，而是提出更多问题，这被称作苏格拉底教育法。
地 理	在东方文化中，老师至今仍享有受人尊敬的崇高地位。
数 学	方程只能算一个最大值，但不能算一个最佳值；作为一位老师，你如何实现职业价值的最大化？
生活方式	我们喜欢惊喜，但不喜欢令人不快的事情，比如考试中那些让人无法回答的问题；对老师知晓全部答案的期待引发了机械式学习；追根溯源与质疑知识、智慧的能力。
社会学	社会变迁中教师角色的演变；在某一特定领域，不需授衔或评级而对领导地位的认可；理解性学习与机械式学习的区别；哲学家在社会中的角色；那些有影响力的人的重要价值——他们影响力的获得并非因为有权力或有金钱，而是因为他们的立场与信仰，因为他们身体力行而不夸夸其谈，因为他们给人以启发。
心理学	我们不管用心学什么，都会逐渐忘记，只有那些与我们的感情相联系的事物才会被记住；挑战的力量让我们不断思考，超越旧识；学生学习的能力。
系统论	社会中的知识、智慧、价值的创立与获得非一日而成，这些品质通常由老师传授，而我们记住了最能启发我们的部分。

情感智慧
Emotional Intelligence

学生一

这名学生起初不明白教授的真实意图，他认为这个任务如果不是一个玩笑，那也太难完成了，因为达到教授的学识水平是不可能的。这名学生更喜欢传统的考试模式，记住教授所教的知识，然后不断重复教授已经知道的一切。通过与学生二的对话，这名学生意识到对于教授来说一些新的考试模式可能是非常有趣的，但是他担心教授不喜欢被自己的学生挑战。最后，他明白了如果教授准备好向自己的学生学习，那么不管是学生还是教授，都会学到更多知识。

学生二

这名学生相信教授是认真的，而且希望接受挑战，学习更多知识。他认为考试是非常枯燥的事情，因为那只是在重复自己和所有人已经知道的旧知识。这名学生愿意启发老师，希望和教授多接触，激发他思考。他信任教授，知道教授在本专业领域是受人尊敬的，并支持这样一种观点：学生应该比自己的老师和父母学到更多知识来推动社会进步。他设想着这样一种循环往复：教授变成大师、超级大师，向学生传递知识，鼓励学生获取新知。反过来，学生也将有机会成长为新一代大师。

艺术
The Arts

我们经常谈及绘画领域的超级大师。列出你心目中的超级绘画大师名单，说一说哪些是他们最好的作品。围坐成一个小组，和同学们一起讨论对比一下各自心目中的超级大师，也许你们会想出一个更全的超级大师名单。你也可以列一些其他领域中的超级大师名单，比如作曲家、音乐家和电影导演。

思维拓展
Systems: Making the Connections

学习是社会的基石，教育通常是一个家庭甚至一个国家最大的一笔投资。每个人都希望去最好的学校向最好的老师学习，然而，学校衡量学生学习能力的方式主要是考试中学生获得的分数。这意味着，他们只专注于一种学习方式。一些学生的学习动机主要来自获得高分，而一些学生则认为分数和学习之间根本没有关系。将考试作为检验方式的另一个问题是学生可能会去作弊，尤其是在一些人数多的班级中。考试中出现越来越多的多项选择题，这意味着机器可以评判试卷，让过程更加客观。随之而来的挑战是要因材施教，不把学习变成标准化的教育。老师过去是社会的核心，而现在已被削弱。此外，老师现在的收入很低，和他们长期的付出不相称。教育系统的运转方式就像是工业化生产，强调专业化与分工，在这里，核心的学习活动是学习核心知识，导致我们只剩下越来越有限的选择。这样的结果是，那些在某一专业学有专攻的学生会感到知识匮乏，需要再获取一些其他学科的文凭。我们的未来依靠那些知道自己有机会比父母学到更多知识的孩子，但是我们应该意识到，这只能通过一定程度的自由来实现，这种自由将创造更多前所未有的新知识、新科学。最好的老师——愿意向自己学生学习的老师，可以引领一些新理念的产生。每个人不仅应该拥有最好的东西，而且还应该有尽己所能做到最好的机会。这就是说，仅有创新能力、知识和智慧转化的方法还不够，还应该有一种全新的师生关系。

动手能力
Capacity to Implement

每个人都可以是你的学生！花时间准备一个你想要讨论的话题，熟练掌握而且能够脱稿讲至少15—20分钟。使用每个人都能听懂的语言，并用一些新的演讲方法，比如个人故事、笑话或惊喜。然后，邀请你的学生来发表评论，分享他们关于这个话题的观点，这时你就该倾听了。最后，记录下你从学生那里学到了什么，有哪些是你从前不知道的、不理解的，有哪些新知识让你为之一振。你所记录的内容是否正确并不重要，重要的是你要理解智慧与知识在人与人之间是如何传播的。

故事灵感来自
This Fable Is Inspired by

魏伯乐教授
Prof. Ernst Ulrich von Weizsäcker

魏伯乐教授是环境、气候和能源政策领域的先驱，他学习了化学和物理学，并获得了生物学博士学位。他是科学界的企业家，是卡塞尔大学的创始校长，也是伍珀塔尔气候、环境和能源研究所的创始主席。2012 年起，他开始担任罗马俱乐部的共同主席。他著有两本介绍如何实现更高的资源利用率的著作《四倍极》和《五倍极》。他还是《私有化的局限》一书的作者之一。作为德国国会议员，他担任议会研究委员会中经济全球化与环境委员会的主席。

图书在版编目（CIP）数据

冈特生态童书.第三辑修订版：全36册：汉英对照 /
（比）冈特·鲍利著；（哥伦）凯瑟琳娜·巴赫绘；
何家振等译.—上海：上海远东出版社,2022
书名原文：Gunter's Fables
ISBN 978-7-5476-1850-9

Ⅰ.①冈… Ⅱ.①冈…②凯…③何… Ⅲ.①生态环
境–环境保护–儿童读物—汉、英 Ⅳ.①X171.1-49

中国版本图书馆CIP数据核字(2022)第163904号
著作权合同登记号图字09-2022-0637号

策　　划　张　蓉
责任编辑　祁东城
封面设计　魏　来　李　廉

冈特生态童书
大师与超级大师
[比]冈特·鲍利　著
[哥伦]凯瑟琳娜·巴赫　绘
田　烁　王菁菁　译

记得要和身边的小朋友分享环保知识哦！
八喜冰淇淋祝你成为环保小使者！